Plant Cells vs
ANIMAL CELLS

Rebecca Woodbury, Ph.D., M.Ed.

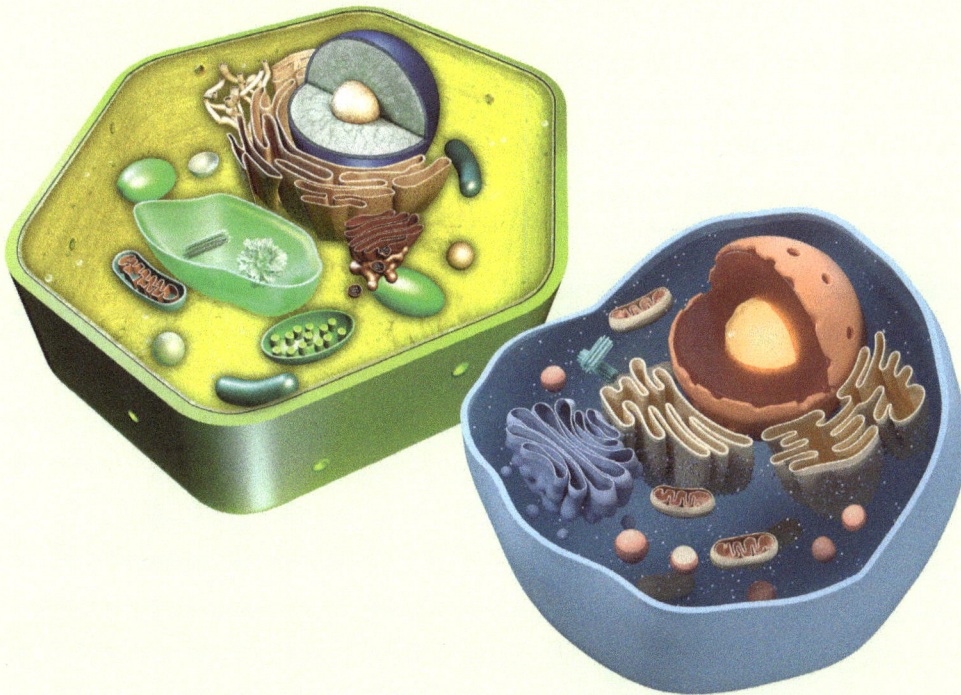

Gravitas Publications Inc.

Plant Cells vs
ANIMAL CELLS

Illustrations: Janet Moneymaker

Plant Cells vs Animal Cells
ISBN 978-1-950415-71-7

Published by Gravitas Publications Inc.
Imprint: Real Science-4-Kids
www.gravitaspublications.com
www.realscience4kids.com

RS4K

Photo credits: Cover, Title Pg, Above, P.9: Top, By Andrea Danti, AdobeStock; Bottom, By Vink Fan, AdobeStock; P.6, 11, 18. By Andrea Danti, AdobeStock; P.7, 10, 19. By Vink Fan, AdobeStock; P.12. Adapted from Illustration by Vink Fan, AdobeStock

Did you know?

Plants and animals have different types of **cells.**

What is a cell?

Turn the page to find out.

Review: THE CELL

- All living things are made of cells.

- A **cell** works like a tiny city.

- Cells do lots of jobs inside living things.

- Cells are made of **atoms** and **molecules.**

Review: ATOMS

- **Atoms** are tiny building blocks that can link together.

- **Atoms** make everything we see, touch, taste, and smell.

Review: MOLECULES

- **Molecules** are made when **atoms link** together.

Plants have **plant cells.**

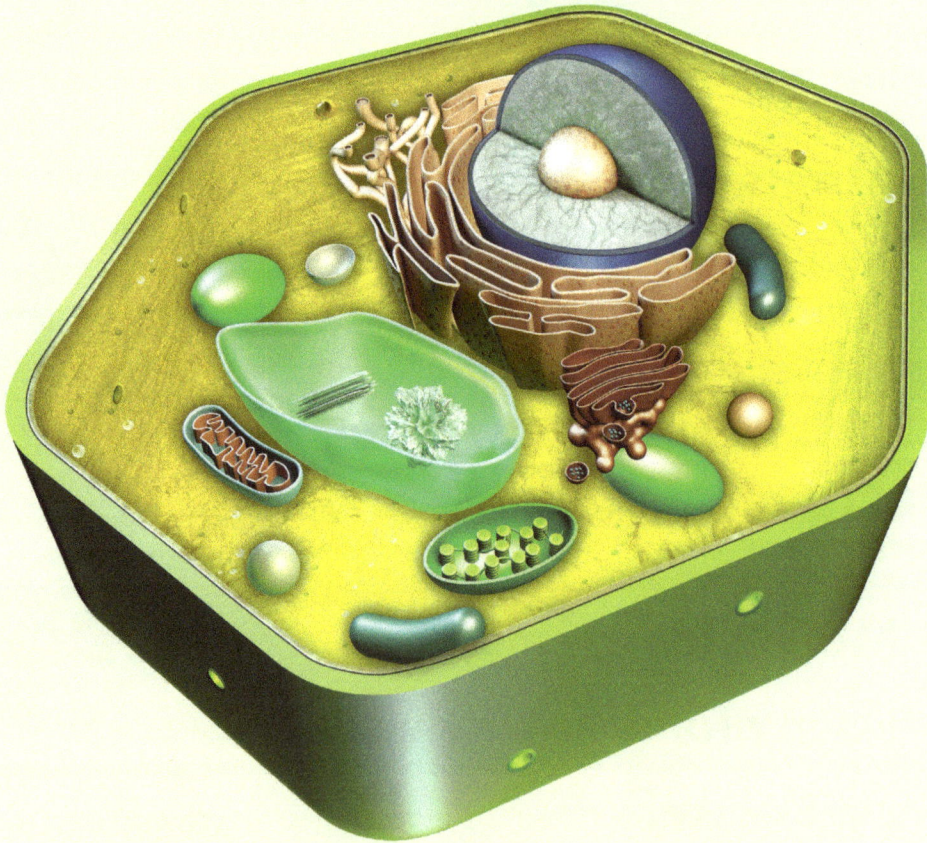

Animals have **animal cells.**

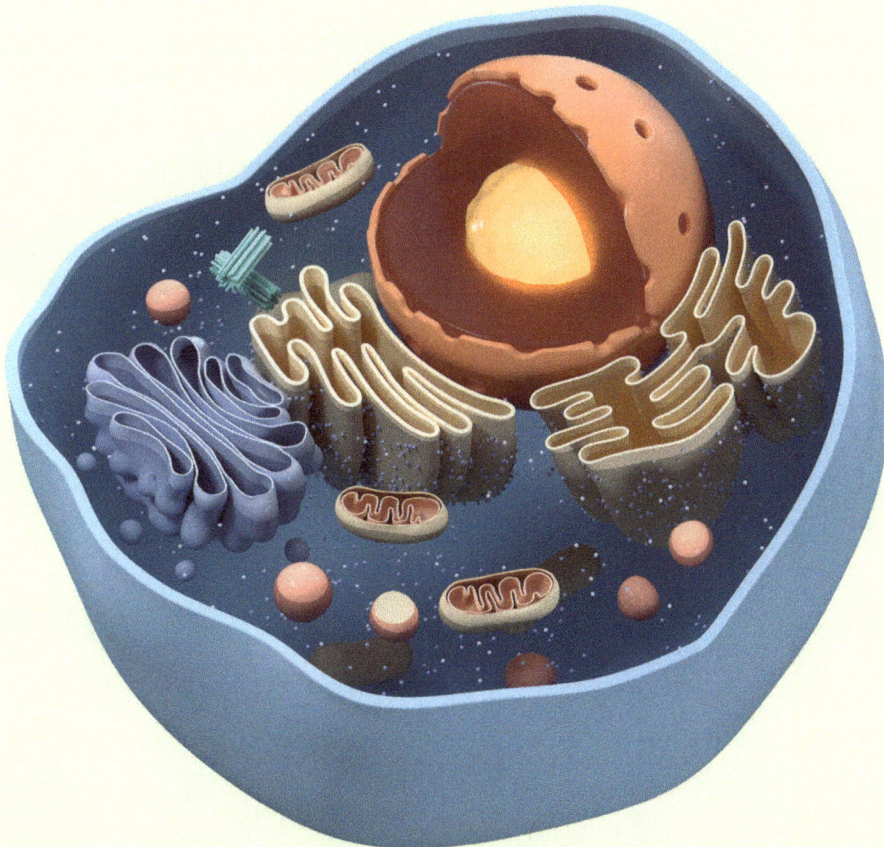

Plant cells and animal cells
are the same in some ways.

Both plant cells and animal
cells are surrounded by a
cell membrane.

Which kind of
cells do we have?

DUH!

Plant Cell

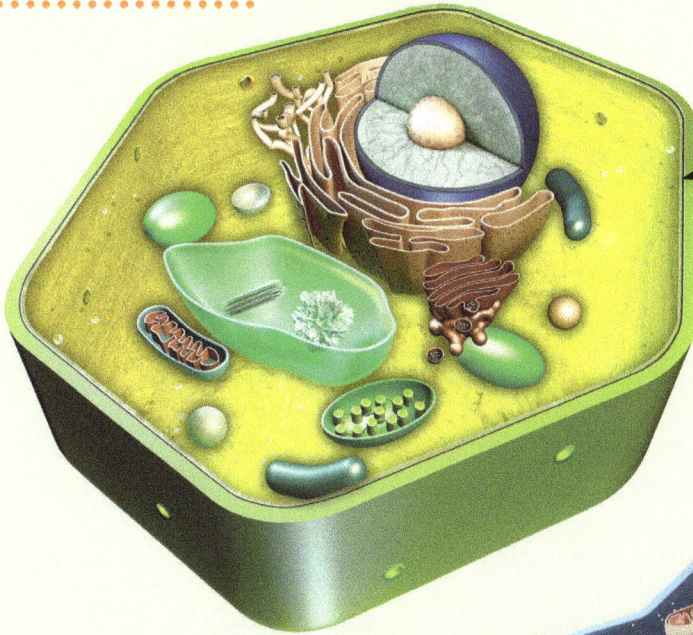

Cell
membrane

Animal Cell

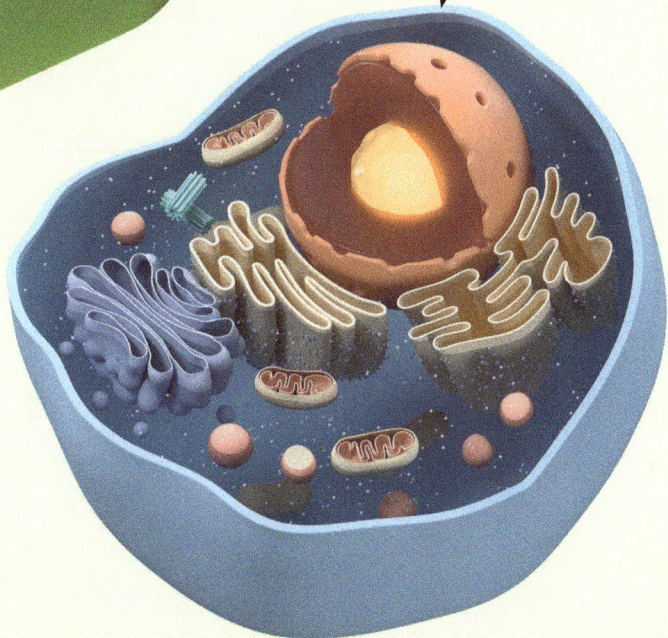

Both animal cells and
plant cells have **organelles.**

Animal Cell

Organelles

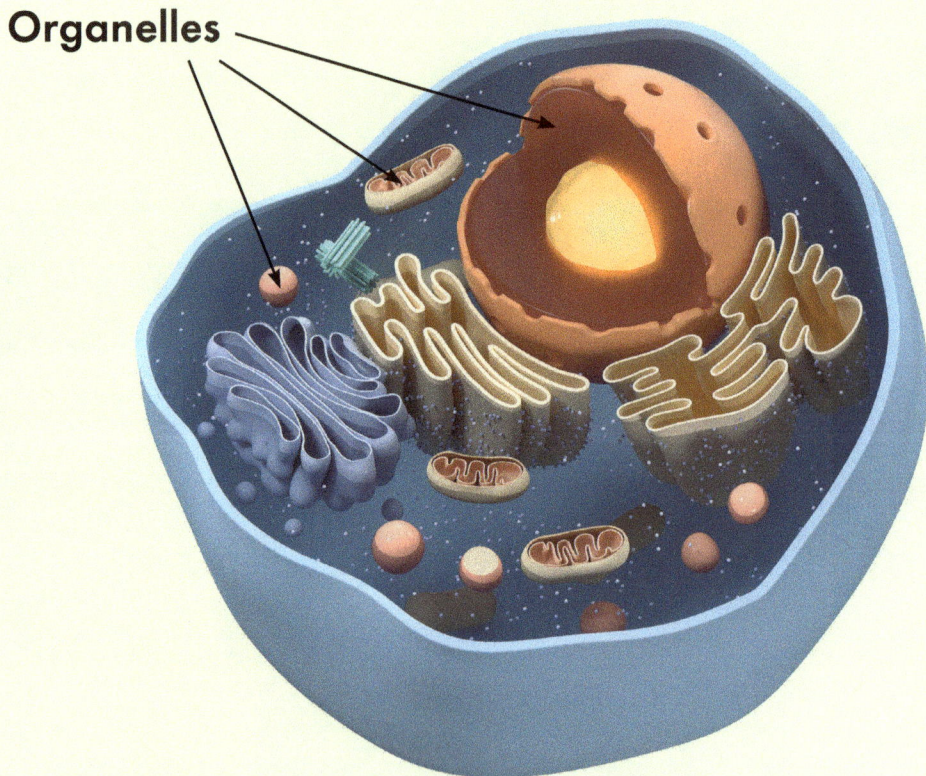

Plant Cell

Organelles

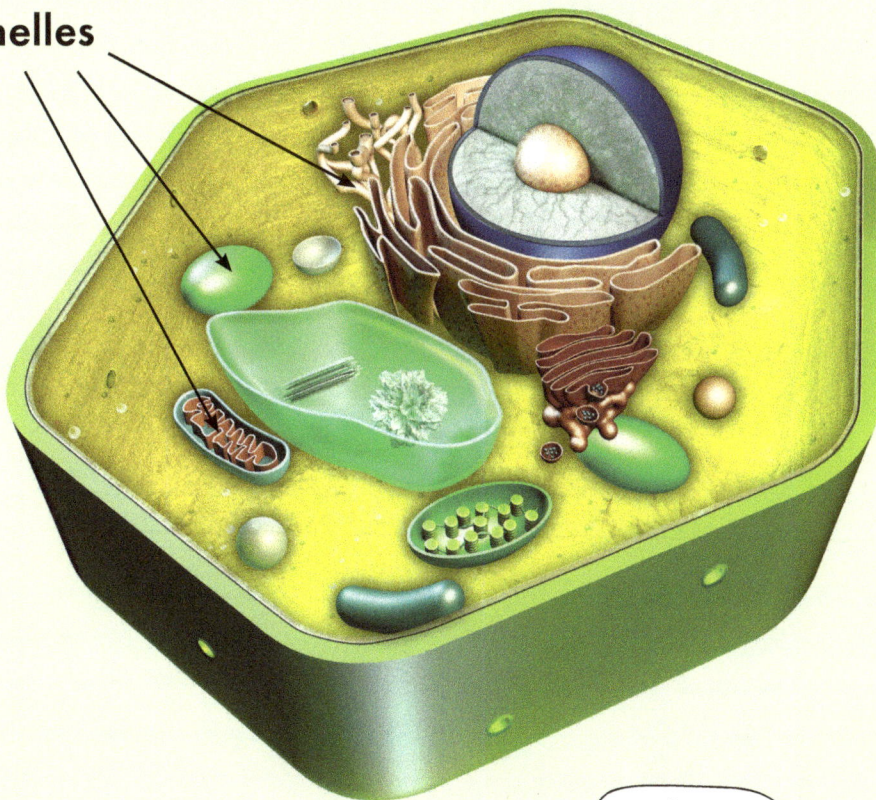

What are organelles?

Look on the next page!

Organelles are small structures inside cells that do certain jobs.

The **nucleus** is the organelle where **DNA** is made and kept.

DNA tells cells what to do.

Nucleus

What is DNA?

Look!

Review: DNA

- **DNA** is a double strand of linked atoms. DNA makes a code that tells cells what to do.

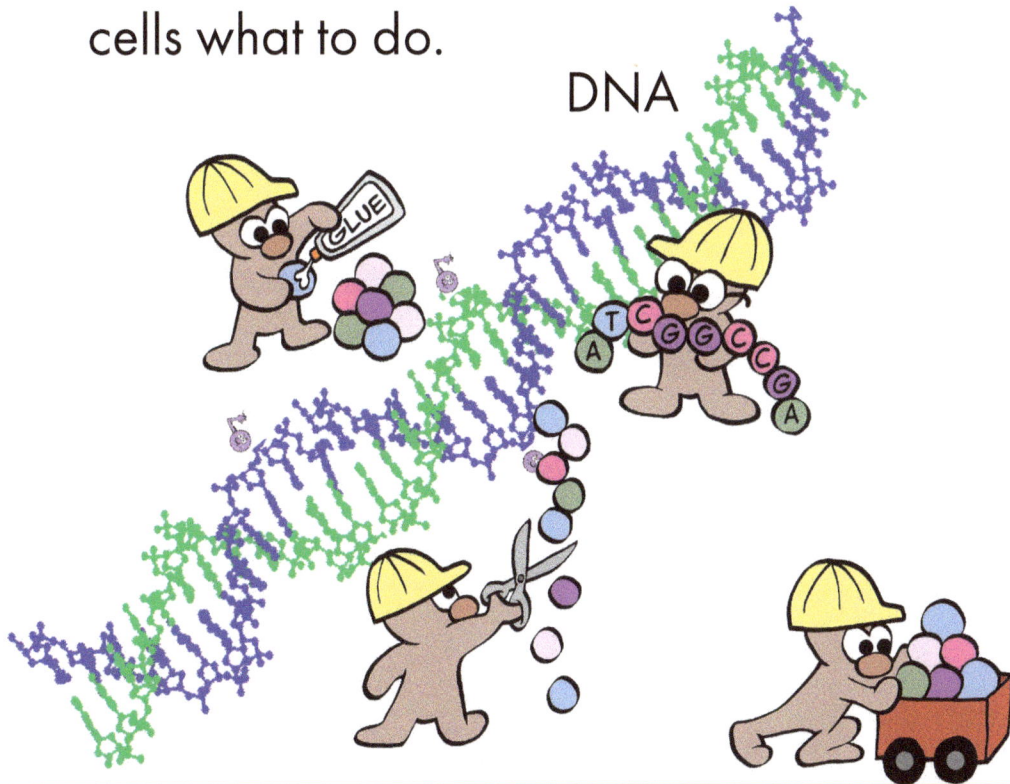

DNA

A **ribosome** is an organelle that makes **proteins** for the cell.

Ribosome

I make proteins.

Review: PROTEINS

- A **protein** is a long chain of atoms.

- Proteins do all the work inside a cell.

- Proteins cut, join, and move molecules.

A **mitochondrion** is an organelle that makes **energy** for the cell.

We make energy.

Review: ENERGY, WORK and FORCE

- **Energy** is the fuel something needs to do **work.**

- **Work** happens when **force** moves an object.

- **Force** is any action that changes the location, shape, or speed of an object.

Do proteins do all the work in a cell?

Yes!

Plant cells and animal cells are also different.

Plant cells have a **cell membrane** AND
a stiff outer **cell wall.**

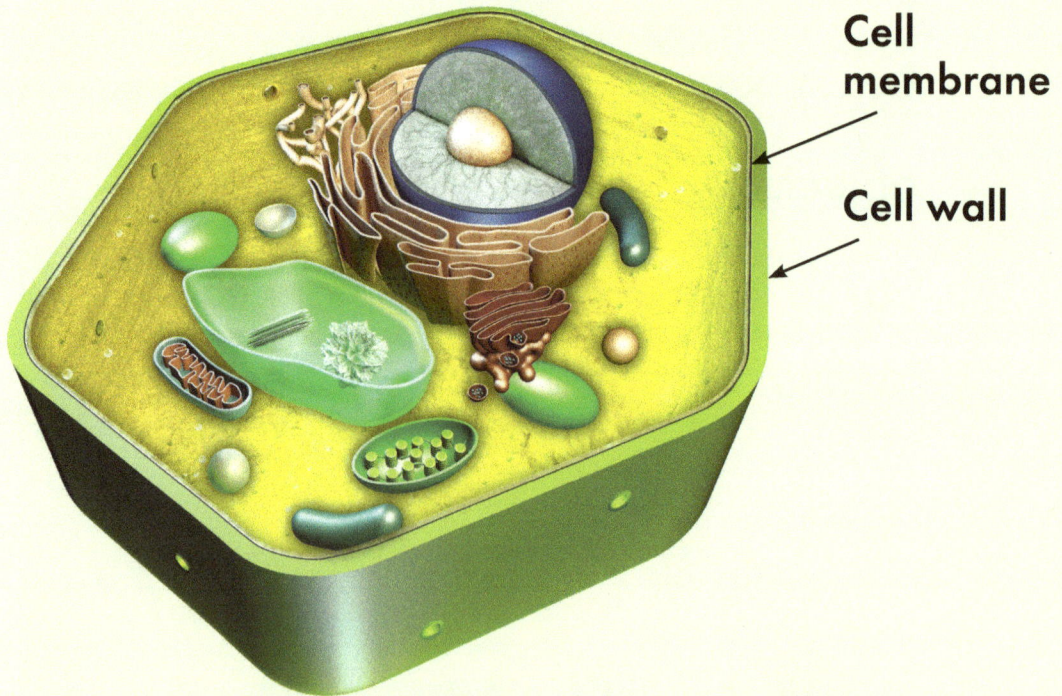

Cell
membrane

Cell wall

Animal cells have only a **cell membrane.**

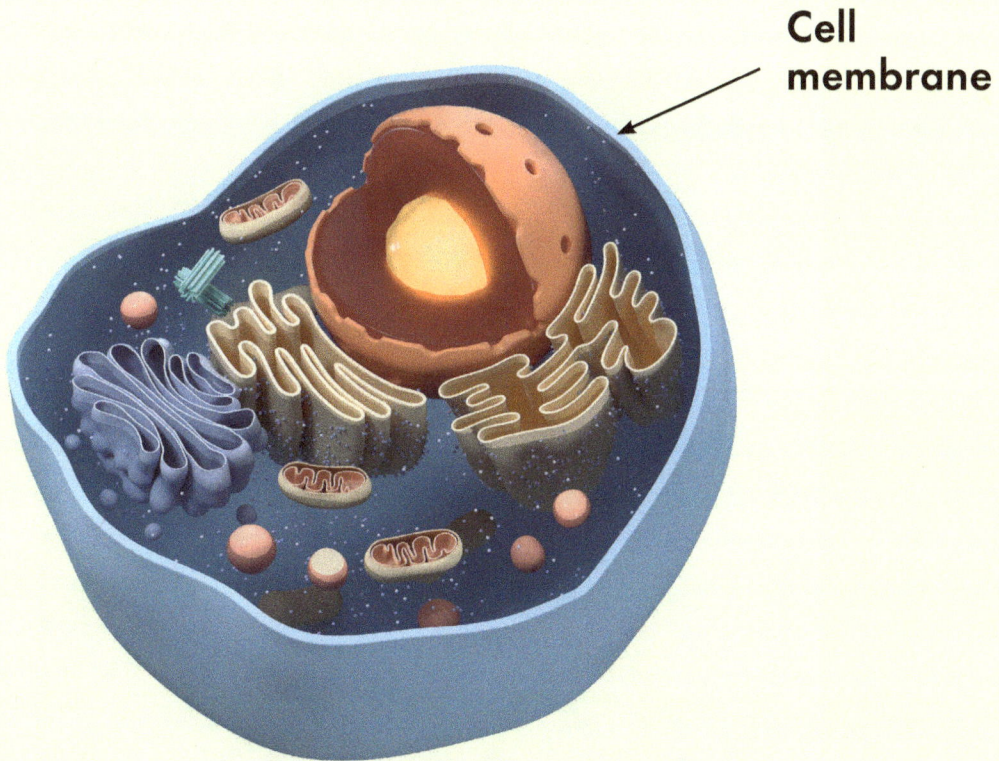

Cell
membrane

Plant cells have **chloroplasts.** A chloroplast is an organelle that catches sunlight to make food.

Animal cells do not have chloroplasts and do not make food from the sun.

Animals must get their food from eating other animals and plants.

Wow! Plant and animal cells are cool!

How to say science words

atom   (AA-tum)

cell   (SEL)

cell membrane   (SEL MEM-brayn)

cell wall   (SEL WAWL)

chloroplast   (KLAW-ruh-plast)

DNA   (DEE - EN - A)

energy   (EH-nuhr-jee)

force   (FAWRS)

mitochondria   (miy-tuh-KAHN-dree-uh) [plural]

mitochondrion   (miy-tuh-KAHN-dree-uhn) [singular]

molecule   (MAH-lih-kyool)

nucleus   (NOO-klee-uhs)

organelle   (AWR-guh-nel)

ribosome   (RIY-buh-sohm)

work   (WERK)

www.ingramcontent.com/pod-product-compliance
Lightning Source LLC
Chambersburg PA
CBHW040151200326
41520CB00028B/7568